RAPS & RHYMES IN MATHS

Compiled by
Ann & Johnny Baker

Heinemann
Portsmouth, NH

Acknowledgements

The authors and publishers thank the following copyright holders for permission to reprint material in this book. If inadvertent errors or omissions have occurred they will be corrected in subsequent editions if they are brought to our notice.

'Five eyes' by Walter de la Mare reproduced by permission of The Literary Trustees of Walter de la Mare and The Society of Authors as their representative.

'Four in a Boat' adapted by permission of Sterling Publishing Co., Inc. from THE BEST SINGING GAMES FOR CHILDREN OF ALL AGES by Edgar S. Bley, © 1957 by Sterling Publishing Co., Inc., 387 Park Ave. S., New York, NY 10016

HEINEMANN EDUCATIONAL BOOKS, INC.
361 Hanover Street Portsmouth, NH 03801
Offices and agents throughout the world

Library of Congress Cataloging-in-Publication Data

Baker, Ann.
Raps & Rhymes & Math/Ann Baker.
p. cm.
ISBN 0-435-08325-2
1. Mathematics — study and teaching (Primary) 2. Mathematical recreations. I. Baker, Johnny. II. Title.00
QA135.5.B2424 1991
372.7 dc20 91-16343
CIP

Published simultaneously in the United States in 1991 by Heinemann and in Australia by Eleanor Curtain Publishing
2 Hazeldon Place
South Yarra 3141

Production by Sylvana Scannapiego, Island Graphics
Designed by Sharon Carr
Typeset by Optima Typesetters
Printed in Australia by Impact Printing Pty Ltd

Contents

Introduction

Rhymes, riddles and stories have been used to pass on the history, beliefs and customs of society from generation to generation. Children learned their lessons at the feet of their elders. Rhymes and stories were the content of those lessons conveying, amongst other concepts, mathematical ideas and understandings. We believe that they still have an important role to play in the learning process and, in particular, in the process of learning mathematics.

The book is a collection of traditional and modern rhymes, riddles and stories with mathematical themes. They can be used to provide a welcome break from more formal activities or they can form the introduction or conclusion of a maths lesson. They also provide a source of enjoyment as the children rap, dramatise or vocalise them. In a previous book, *Mathematics in Process* we emphasised the power of rhymes and stories, showing how they blend with a 'whole' approach to teaching mathematics, involving music, arts, crafts and social sciences. And of course they can also be used to provide openings for mathematical investigations.

Share the rhymes with the children by writing them on the board, or on a large sheet of paper. Read them aloud, with the class following silently. Then it's up to you and the children to try a variety of readings, improvise performances, and model and explore the mathematical ideas.

This book is a starting point for using rhymes, riddles and stories in the maths lesson. Rather than give activities or suggestions for each rhyme, we have provided an outline of what happened when we used one particular rhyme. Other similar rhymes then follow so that you and your children can bring your own stamp of individuality and creativity to them.

Above all, this book emphasises enjoyment. Let the children have fun with these rhymes, riddles and stories and follow the leads that they present or that the children find interesting.

LOWER PRIMARY

5 — 7 YEARS

Counting out and Counting on

Three young rats

All the children click (*) to the rhythm while introducing counting on and counting down into this traditional rhyme.

Three young rats with black felt hats,
*One, *two, *three.
Three young ducks with white straw flats,
*Four, *five, *six.
Three young dogs with curling tails,
*Seven, *eight, *nine.
Three young cats with demi-veils,
*Ten, *eleven, *twelve.
Went out to walk with three young pigs,
*Thirteen, *fourteen, *fifteen.
In satin vests and sorrel wigs.
But suddenly it chanced to rain
And so they all went home again.
Fifteen, fourteen, thirteen,
Twelve, eleven, ten,
Nine, eight, seven,
Six, five, four,
Three, two, one.

(All clap in time with the counting, increasing the tempo as the numbers get smaller)

Activities

Counting

After we had introduced this rhyme, we grouped the children in threes to dramatise the second reading. This promoted a wide range of counting activities at a variety of levels.

- Some children had to solve the problem of including themselves in the count, as their first move was for one child to collect three others resulting in groups of four.
- Some children asked 'How many is that altogether?' and counted in ones before trying to count in threes.
- The question most asked after the second reading was 'Can we all be in it?' which began a spontaneous investigation into how many groups of three the class could make, or how many should be in each group to make the rhyme work. This resulted in two improvisations on the rhyme, the first had five in each group of animals, whereas the second had two extra groups of animals added with the leftover child 'reading' the rhyme.

Dramatisation and matching

Props were made so that the dramatisation could be embellished. Since props will go astray, one-to-one matching became a feature of this activity as questions arose, such as:

'Is there a hat for each duck?'

'How many vests do we have to find?'

By varying the number of participants in the groups, these questions become more common and more varied. Each child in the class chose an animal to draw, pasted it onto a paddlepop stick and made the appropriate props for it. As the children played with these figures they asked questions such as:

'Are there more pigs than ducks? How many more?'

'Does each animal have a partner?'

'What if the ducks stayed home?'

Combinations

As the children re-enacted the rhyme with the puppets they began to swap the props around making different combinations, and so questions were asked, for example:

> 'If we used ducks and pigs and their props how many different outfits are there?'

Improvisation

Improvising on the rhyme allowed many possibilities other than those that the children had already generated, for example:

- the introduction of other creatures and categories (e.g. big cats, reptiles)
- designing different items of apparel, looking at the history of fashion or costumes of other countries
- extending the range of numbers worked with in terms of either the number of groups, or the number in each group, or both.

Recording

The children illustrated their own favourite line of the rhyme and wrote its corresponding number and number name. They also illustrated one line of their improvised rhymes. The cut-outs were used to make graphs for making comparisons.

Some children recorded number stories to match their improvised rhymes,

> 'We had six groups of big cats, with four in each group. Twenty-four altogether.'

The children also wrote about what they found as they carried out their investigations and what strategies they used. These were shared and compared with others.

Other counting on rhymes

The following rhymes also have counting themes. We have given a few starter ideas for them but we suggest you share them with your children and see what you and they together come up with. As the children dramatise and improvise on the rhymes, mathematical ideas and questions that the children will want to explore will no doubt arise. These will be far more memorable and relevant than any we could generate, and the level of complexity will be tailored to the needs of each individual learner.

Baa Baa Black Sheep

Use props, e.g. cut-out paper sacks to dramatise this and extend the counting to match the number in the group. Try this for other numbers, e.g. six, so that they have two sacks each, or one, two and three respectively. The children can then explore numbers of their choice.

Baa Baa Black Sheep
Have you any wool?
Yes Sir, Yes Sir,
Three bags full.
One for the Master,
One for the Dame,
And one for the little boy,
Who lives down the lane.

Magpies

A group of eight children click to the rhythm while standing up and sitting down as follows:

We saw eight magpies in a tree,	(All standing)
Two for you and six for me.	(Two, then six sit down)
One for sorrow, two for mirth,	(1st then 2nd stand)
Three for a wedding, four for a birth,	(3rd then 4th stand)
Five for England, six for France,	(5th then 6th stand)
Seven for a fiddler, eight for a dance.	(7th then 8th stand)

Illustrating this rhyme provides opportunities for counting and labelling as well as illustrating and exploring number facts, 2+6=8, or improvising.

> 'I saw eight magpies in a tree, three for you and five for me.'
>
> 'I saw ten magpies in a tree . . . '

This rhyme makes a good 'dipping' or choosing rhyme too, and may lead to investigations of where to begin the rhyme or where to stand to make sure that you are the last one 'in'.

Here is the beehive

One child faces the rest of the group and leads the chant:

Child 1	**Here is the beehive**
All	**Where are the bees?**
Child 1	**Hidden away where nobody sees.**
	Here they come now,
All	**1, 2, 3, 4, 5**
	Buzz, buzz, buzz, buzz, buzz.

For young children, matching the number of buzzes to the number of bees is quite a challenge. By varying the number of bees the rhythm of different counting patterns can be explored.

1, 2, **3**	1, 2, 3, **4**	1, 2, 3, 4, **5**
4, 5, **6**	5, 6, 7, **8**	6, 7, 8, 9, **10**

Counting patterns

There are many rhymes and jingles that emphasise counting in twos. Any of these rhymes can be illustrated, dramatised or improvised on to provide an introduction to counting patterns.

2, 4, 6, 8

2, 4, 6, 8
Mary at the cottage gate,
Eating cherries off a plate.

2, 4, 6, 8,
Bog in don't wait.

2, 4, 6, 8,
Who do we appreciate?

One, two, buckle my shoe

Two groups chant the rhyme alternately, and the second group claps (•) and clicks (*) the rhythm:

Group 1	One, two,
Group 2	Buckle my shoe.
Group 1	Three, four,
Group 2	Knock at the door.
Group 1	Five, six,
Group 2	Pick up sticks.
Group 1	Seven, eight,
Group 2	Lay them straight.
Group 1	Nine, ten,
Group 2	A big fat hen.
Group 1	Eleven, twelve,
Group 2	Dig and delve.
Group 1	Thirteen, fourteen,
Group 2	Maids a-courting.
Group 1	Fifteen, sixteen,
Group 2	Maids in the kitchen.
Group 1	Seventeen, eighteen,
Group 2	Maids in waiting.
Group 1	Nineteen, twenty,
Group 2	My plate's empty.

Once I caught a fish alive

One child says the rhyme, while the rest of the group asks the questions.

Group	**One two three four five.**
Child 1	**Once I caught a fish alive.**
Group	**Six seven eight nine ten.**
Child 1	**Then I let it go again.**
Group	**Why did you let it go?**
Child 1	**Because it bit my finger so!**
Group	**Which finger did it bite?**
Child 1	**This little finger on my right.**

Counting down

Five little mice

All whisper the first two lines and wriggle the fingers of the left hand. All shout the next line and the right hand becomes the cat. All say the next line as the cat jumps on a mouse.

Five little mice went out to play,
Looking for crumbs to eat on the way.
OUT jumped a pussy cat bold and black.
And four little mice came scampering back.

Four little mice went out to play,
Looking for crumbs to eat on the way.
OUT jumped a pussy cat bold and black.
And three little mice came scampering back.

Three little mice went out to play,
Looking for crumbs to eat on the way.
OUT jumped a pussy cat bold and black.
And two little mice came scampering back.

Two little mice went out to play,
Looking for crumbs to eat on the way.
OUT jumped a pussy cat bold and black.
And one little mouse came scampering back.

One little mouse went out to play,
Looking for crumbs to eat on the way.
OUT jumped a pussy cat bold and black.
And no little mice came scampering back.

An alternative is for the children to click (*) and clap (•).

Five little mice went out to play,
Looking for crumbs to eat on the way.
OUT jumped a pussy cat bold and black.
And four little mice came scampering back

Activities

Counting

We encouraged the children to predict how many mice would be left each time so that they became familiar with the language and recall of 'one less'. They later expressed what happened in each verse as number stories, for example:

> 'There were three mice, and one got caught.
> Now there are only two mice left.'

Improvisation

Improvising on the rhyme allowed many possibilities:
- discussion of creatures and their predators (e.g. lizards and flies)
- inventing, drawing and naming fantasy creatures
- extending the range of numbers
- varying the number of creatures who get caught each time.

This led to investigations suggested by the children:

> 'If we want the cat to catch three mice each time, what numbers could we start with so that all the mice get caught?'

Recording

Younger children illustrated their own versions of 'Five little mice', while the older children illustrated and wrote number stories to match each verse of their improvised rhyme. Some children wrote number problems about their rhymes, for example:

> There were 7 baby dinosaurs.
> Tyrannosaurus has eaten 3.
> How many are left?

Other counting down rhymes

The following rhymes each have counting down themes and lend themselves to improvisation, dramatisation, modelling and illustration.

Five currant buns

This is played with fingers.

Five currant buns in a baker's shop,
Round and fat with sugar on top.
Along came a boy/girl with a dollar one day,
Bought a currant bun and took it away.

Four currant buns, etc.

Modelling this rhyme by paying for each bun and taking a bun away each time gives visual images of 'one less than'.

Five little leaves

Five children stand up and combine clicks (*) with actions.

*Five little *leaves so *bright and *gay
Were *dancing *about on a *tree one *day.
The *wind came *blowing *through the *town
(All sway)
Oooooo . . . Oooooo . . .
(One child falls down)
*One little *leaf came *tumbling *down.

Four little leaves so bright and gay
Were dancing about on a tree one day.
The wind came blowing through the town
Oooooo . . . Oooooo . . .
One little leaf came tumbling down.

Three little leaves so bright and gay
Were dancing about on a tree one day.
The wind came blowing through the town
Oooooo . . . Oooooo . . .
One little leaf came tumbling down.

Two little leaves so bright and gay
Were dancing about on a tree one day.
The wind came blowing through the town
Oooooo . . . Oooooo . . .
One little leaf came tumbling down.

One little leaf so bright and gay
Was dancing about on a tree one day.
The wind came blowing through the town
Oooooo . . . Oooooo . . .
The last little leaf came tumbling down.

We improvised versions of the rhyme so that the children could model the actions and predict then check how many leaves would be left each time.

Ten fat sausages

Ten fat sausages sitting in the pan,
One went Pop and another went Bang.

(Show ten fingers)
(Clap hands loudly)

Eight fat sausages sitting in the pan,
One went Pop and another went Bang.

(Adjust the number of fingers held up in each verse)

Six fat sausages sitting in the pan,
One went Pop and another went Bang.

Four fat sausages sitting in the pan,
One went Pop and another went Bang.

Two fat sausages sitting in the pan,
One went Pop and another went Bang.

No fat sausages sitting in the pan.

This rhyme introduces counting down in twos and can be easily extended to counting down in other groupings or counting beyond 10, e.g.

> 100 fat sausages sitting in the pan,
> 10 went Pop and 10 went Bang.

When the children chose their own starting numbers and counting down patterns, some interesting adjustments had to be made to the final verse to result in none being left.

Five little pussy cats

Five children play the cats, half the rest of the class says the rhyme, pausing to let the remaining children say the numbers.

Five little pussy cats playing near the door;
One ran and hid inside and then there were **four**.
Four little pussy cats underneath a tree;
One heard a dog bark and then there were **three**.
Three little pussy cats thinking what to do;
One saw a little bird and then there were **two**.
Two little pussy cats sitting in the sun;
One ran to catch his tail and then there was **one**.
One little pussy cat looking for some fun;
He saw a butterfly and then there were **none**.

Acting out

'Roll over' and 'Five little sparrows' are great for acting out and can be varied to include the whole group or class as well as different counting down patterns.

Roll over

The children dramatise the rhyme, with the smallest saying 'Roll over' and the rest rolling away, one by one. A special effect, e.g. a drum roll, represents a child falling out of bed.

All **There were ten in the bed and the little one said,**
Child **'Roll over, roll over.'**
All **So they all rolled over and one fell out.** (Special effect)

All **There were nine in the bed and the little one said,**
Child **'Roll over, roll over.'**
All **So they all rolled over and one fell out.** (Special effect)

All **There were eight in the bed and the little one said,**
Child **'Roll over, roll over.'**
All **So they all rolled over and one fell out.** (Special effect)

All **There were seven in the bed and the little one said,**
Child **'Roll over, roll over.'**
All **So they all rolled over and one fell out.** (Special effect)

All **There were six in the bed and the little one said,**
Child **'Roll over, roll over.'**
All **So they all rolled over and one fell out.** (Special effect)

All **There were five in the bed and the little one said,**
Child **'Roll over, roll over.'**
All **So they all rolled over and one fell out.** (Special effect)

All There were four in the bed and the little one said,
Child 'Roll over, roll over.'
All So they all rolled over and one fell out. (Special effect)

All There were three in the bed and the little one said,
Child 'Roll over, roll over.'
All So they all rolled over and one fell out. (Special effect)

All There were two in the bed and the little one said,
Child 'Roll over, roll over.'
All So they all rolled over and one fell out. (Special effect)

All There was one in the bed and the little one said,
Child 'Aaah . . . Now I can get some sleep!'

Five little sparrows

Use the fingers of one hand to indicate the number of sparrows and match the action of the rhyme.

All **Five little sparrows sitting in a row.**
Child 1 **One said, 'Cheep, cheep, I must go!'**
All **One little, two little, three little, four little,**
Five little sparrows — Oh.
(One little sparrow flies away)

All **Four little sparrows sitting in a row.**
Child 2 **One said, 'Cheep, cheep, I must go!'**
All **One little, two little, three little,**
Four little sparrows — Oh.
(One little sparrow flies away)

All **Three little sparrows sitting in a row.**
Child 3 **One said, 'Cheep, cheep, I must go!'**
All **One little, two little,**
Three little sparrows — Oh.
(One little sparrow flies away)

All **Two little sparrows sitting in a row.**
Child 4 **One said, 'Cheep, cheep, I must go!'**
All **One little,**
Two little sparrows — Oh.
(One little sparrow flies away)

All **One little sparrow left in the row.**
Child 5 **Said, 'Oh, dearie me what shall I do?'**
All **One little, two little, three little, four little, five little, —**
Child 5 **'Cheep! I'll fly away too.'**

Number facts

Swinging on a gate

Click (*) and clap (•) to the rhythm of the rhyme.

Swinging on a gate, swinging on a gate,
Seven little sisters and a brother makes eight.
Seven pretty pinafores and one bow tie,
Fourteen pigtails and one black eye.

Swinging on a gate, swinging on a gate,
Seven little sisters and a brother makes eight.
Seven pretty pinafores and one little cap,
Eight books for school caught up in a strap.

Swinging on a gate, swinging on a gate,
Seven little sisters and a brother makes eight.
The school bell rings and off they go,
Eight little children all in a row.

Activities

Free response

When we read this rhyme to the children they asked:

'Are there really fourteen pigtails on seven little girls?'

'What if some girls only had one pigtail or if they had none?'

'Some boys have a little pigtail too, how many then?'

The size of the family caused a great deal of discussion with the children comparing family size (some realising that they also had to include themselves in the count).

Improvising

The children didn't really feel that boys and girls needed to be categorised in the way that they were. As one little girl put it,

> 'I got a black eye once.'

So in groups of varying sizes the children chose characteristics that they thought were relevant: missing teeth, eye colour, hair colour, uniform or no uniform. Sisters and brothers were replaced with 'children', 'kids', 'people'. The new rhymes were illustrated and made into a class book. Some children made up rhymes about the whole class, extending the range of numbers to suit their level of confidence. One group used bits of wool as ribbons and moved into 2s, 3s, 4s and 5s counting patterns.

Recording

Some children simply illustrated their rhymes to show the combinations while some wrote their rhymes and added number sentences too.

Other number fact rhymes

The following rhymes can all be used to present or introduce different number facts. The children can model the actions with objects and record, in their own way, what happens each time. Frequently children use objects to model or investigate number facts and then record what they have done. They are less frequently invited to work the other way round, that is from a number story or sentence to the concrete representation of it. These rhymes provide ideal opportunities for this to happen.

1 and 1 are 2

The class chants the number facts and the teacher answers with the rhyme.

Class 1 and 1 are 2
Teacher **That's for me and you.**

Class **2 and 2 are 4**
Teacher **That's a couple more.**

Class **3 and 3 are 6**
Teacher **Barley-sugar sticks.**

Class **4 and 4 are 8**
Teacher **Tumblers at the gate.**

Class **5 and 5 are 10**
Teacher **Bluff seafaring men.**

Class **6 and 6 are 12**
Teacher **Garden lads who delve.**

Class **7 and 7 are 14**
Teacher **Young men bent on sporting.**

Class **8 and 8 are 16**
Teacher **Pills the doctor's mixing.**

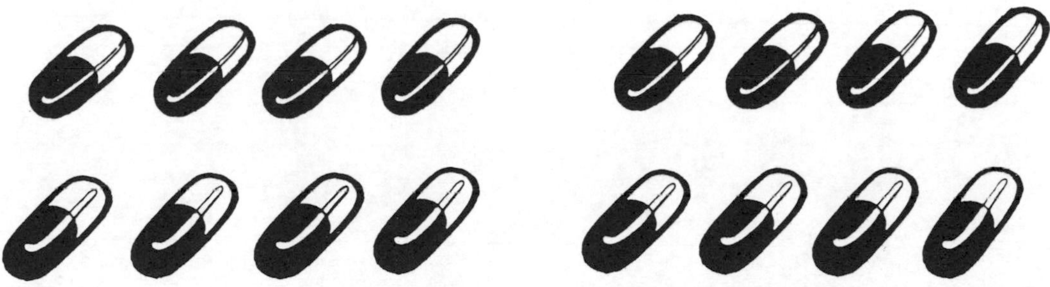

Class 	**9 and 9 are 18**
Teacher **Passengers kept waiting.**

Class **10 and 10 are 20**
Teacher **Roses, pleasant plenty!**

Class **11 and 11 are 22**
Teacher **Sums for brother George to do.**

Class **12 and 12 are 24**
Teacher **Pretty pictures, and no more.**

Improvising

The children improvised on this rhyme using doubles 1 to 6 and some used doubles 13 to 20. One group went on to trebles,

'7+7+7 are 21
Young boys having fun.
8+8+8 are 24
There's the doctor at the door.'

Recording

The children's groupings for the rhyme were pasted onto paper as a permanent class display, along with the rhyme and number sentences. The children were fascinated by the variations and soon wanted to explore each other's ideas.

The improvised rhymes were illustrated and made into a class book.

Spiders

The children each make a spider and use them to dramatise the rhyme.

Ten little spiders went out one day,
Out on their spider's web to play.
Down flew a blackbird and gobbled up three,
Seven little spiders spun back to their tree.

Seven little spiders went out one day,
Out on their spider's web to play.
Along came a duster and whisked away four,
Three little spiders dropped back to the floor.

Three little spiders went out one day,
Out on their spider's web to play.
The wind came up and blew and blew,
One blew away and that left two.

Two little spiders went out one day,
Out on their spider's web to play.
Two little spiders swinging in the sun,
Swung off their web and then there were none.

Improvising

After the children had modelled and recorded or restated what they have done they improvised on this rhyme in a variety of ways, for example they:

- changed the number that disappears each time
- extended the quantity worked with
- included some going off with some coming back
- wrote number stories for others to model the actions to, e.g.

> 'Three went away, then two, then one, then the last four went away. How many were there to start with?'

Ten little squirrels

Class	**Ten little squirrels sat on a tree**	(Show ten fingers)
Child 1&2	**The first two said, 'Why, what do we see?'**	(Hold up thumbs)
Child 3&4	**The next two said, 'A man with a gun.'**	(Hold up fore fingers)
Child 5&6	**The next two said, 'Let's run, let's run.'**	(Hold up middle fingers)
Child 7&8	**The next two said, 'Let's hide in the shade.'**	(Hold up ring fingers)
Child 9&10	**The next two said, 'Why, we're not afraid,'**	(Hold up the little fingers)
Class	**But 'Bang' went the gun, And away they all ran.**	(Clap loudly and hide all fingers)

Explaining

Careful reading of this rhyme reveals that perhaps not all the squirrels left before the gun went off. The children investigated this possibility as they dramatised, modelled or illustrated each line of the verse. Since a variety of interpretations were made, we asked the children to explain their answers to each other.

Countdown

Ten children say the lines of the rhyme; the rest of the class say 'Ooo' at the end of each line, except for the last line when all shout 'Boo'.

There are ten ghosts in the pantry, Ooo
There are nine upon the stairs, Ooo
There are eight ghosts in the attic, Ooo
There are seven on the chairs, Ooo

There are six within the kitchen,   Ooo
There are five along the hall,   Ooo
There are four upon the ceiling,   Ooo
There are three upon the wall,   Ooo
There are two ghosts on the carpet,   Ooo
There is one ghost right behind me,   Ooo
Who is oh so quiet . . .

BOO.

After sharing this rhyme we asked the children to estimate how many ghosts were in the rhyme altogether. The children make ghosts from tissues and rubber bands so that they could use them if they wanted to check how many were in the rhyme. One group set the ghosts out systematically in rows:

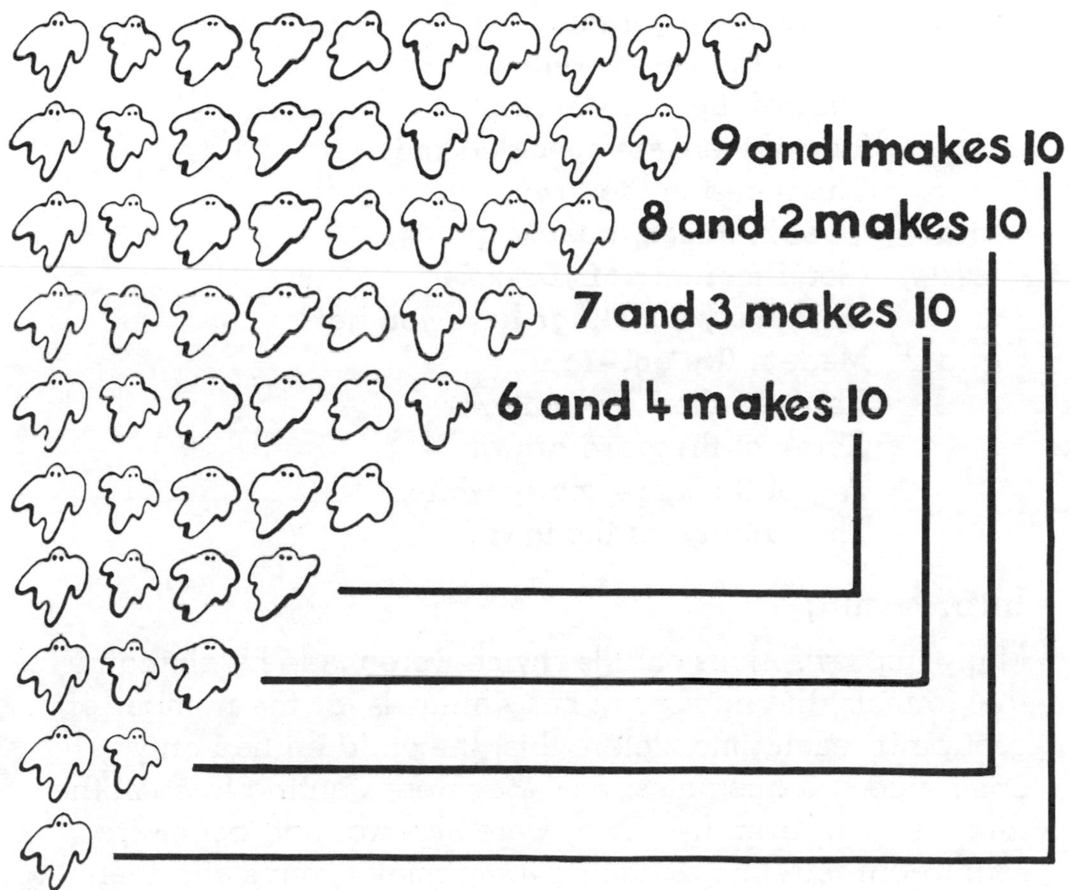
9 and 1 makes 10
8 and 2 makes 10
7 and 3 makes 10
6 and 4 makes 10

They observed the triangular pattern:

'There's one less each time.'

and then saw how moving the one to the row with nine made another ten, and moving the two to the row with eight also made ten, and so on. This led to them trying to count in tens. Their method was demonstrated to the rest of the class who also tried it out and then a class poster was made to show what they had done.

Families

Two children answer the questions that the class asks:

Child 1 **Cluck, cluck, cluck, cluck,**
Class **Good morning Mrs Hen.**
How many chickens have you got?
Child 1 **Madam I've got ten.**
Four of them are yellow,
Four of them are brown,
Two of them are speckled red,
The nicest in the town.
Child 2 **Quack, quack, quack, quack,**
Class **Good morning Mr Drake.**
How many ducklings have you got?
Child 2 **Madam I've got eight.**
Three of them are yellow,
Three of them are brown,
Two of them are snowy white,
The prettiest in the town.

Improvising

Many improvisations on this rhyme were made by changing the animals, the number of baby animals, or the number of groupings each time. Before this, the children had only combined two quantities; now they were wanting to combine three, four or even five. They were also working backwards, setting out ten objects, taking away small groups and then reversing the process to double check.

Construction

Build a house with five bricks

All use gestures to model the rhyme.

Build a house with five bricks,
One, two, three, four, five.
(Use clenched fists for bricks)
Put a roof on top
(Raise both arms above head with fingers touching)
And a chimney too,
(Straighten arms)
Where the wind blows through.
Whoo Whoo.
(Blow hard or whistle, wobble and fall down)

Activities

Visualising

As we read and dramatised this rhyme we asked the children to visualise what the house looked like and some went on to try to draw the house. This led to discussion about the differences in the houses, some were single storey style in the children's minds and some were two to five storey buildings of various designs.

3D building

From these discussions the children decided to explore what sort of houses they could build with five bricks. The variety of designs prompted questions like,

> 'Are these houses really the same or are they different?'
>
> 'Suppose we change one brick each time?'

Some children used cardboard to make roofs which gave rise to questions such as,

> 'How do you make a flat piece of card into a roof?'
>
> 'Which house needs the smallest/largest roof?'

At first the roofs were simply constructed by folding card in half and this left gaps where the wind could blow through. Some of the older children soon wanted 'proper roofs' and so experimented with things such as corners, end pieces and different levels.

Improvising

The children improvised on this rhyme by changing the house to a castle, bridge, boat and so on, and by changing the number of bricks worked with, e.g.

> Build a boat with twelve bricks,
> One, two, three on the bottom
> Four, five, six and seven for the middle,
> Eight, nine, ten, eleven, twelve make the deck.
> Now find a cylinder for the funnel . . .
> Whoo whoo whoo!

Recording

We suggested to the older children that they try drawing their houses or plans for their houses so that others could come along and build them again later. Some of the improvised rhymes were recorded for the younger children to listen and count along to.

Other construction rhymes

The following rhymes can be used to generate other construction activities using a variety of construction materials.

Build a house

Three groups take it in turns to sing a line of the song. All join in for the ending.

Group 1	**Build a house with a floor, with a floor, with a floor.**
Group 2	**Build a house with a wall, with a wall, with a wall.**
Group 3	**Build a house with a roof, with a roof, with a roof.**
All	**Ooooh!**
	There's no house any more.

We introduced this rhyme using a pack of playing cards. As the rhyme unfolded the children built the house to see how far they could get before their house fell down. We then challenged the children to,

> 'Find a strong way of building with cards.'
>
> 'Build the largest house possible.'

London Bridge is falling down

Two groups take it in turns to sing a verse of the song.

Group 1 **London Bridge is falling down,**
Falling down, falling down,
London Bridge is falling down,
My fair lady.

Group 2 **Build it up with wood and clay,**
Wood and clay, wood and clay,
Build it up with wood and clay,
My fair lady.

Group 1 **Wood and clay will wash away,**
Wash away, wash away,
Wood and clay will wash away,
My fair lady.

Group 2 **Build it up with bricks and mortar,**
. . .

Group 1 **Bricks and mortar will not stay,**
. . .

Group 2 **Build it up with iron and steel,**
. . .

Group 1 **Iron and steel will bend and bow,**
. . .

Group 2 **Build it up with silver and gold,**
. . .

Group 1 **Silver and gold will be stolen away,**
. . .

Group 2 **Set a man to watch all night,**
. . .

Group 1 **Suppose the man should fall asleep?**
. . .

Group 2 **Give him a pipe to smoke all night,**
. . .

'London Bridge is falling down' can generate exploration of strength and shape too as the children use a variety of materials, card, newspaper, peas and cocktail sticks, blocks, Lego, constructo straws, and so on to make a bridge.

We used this rhyme before setting a group of older children the challenges:

'Make a bridge to span 15/30/50 cm.'

'Make a bridge strong enough to support the weight of this book.'

'Make a bridge that opens.'

Build it up

The children clap or build with blocks as they sing the song.

Build it up, build it up
Build it high.
Build it high, high, high
Into the sky.

The children enjoyed building with blocks to the rhythm of 'Build it up' and loved the moment of the crash as their wobbly constructions fell down. After the first few hilarious experiences the children began to play in earnest as they tried to find out just how high they could build, asking questions such as:

'Which block is best to start with?'

'What if we have a bigger base?'

'How can we balance it?'

'Can we use all the blocks in one building?'

Measuring

Hands

One hand, two hands, three hands, four!
Four hands wide is the width of the door.
One, two, three, four, five, six, seven!
Seven hands tall is the height of Kevin.
Two hands, four hands, six hands, eight!
Eight hands long is the length of Kate.
Use your hands I'm sure you're able,
And find out the height, length, width of a table.

Activities

Informal measurement

We read this rhyme to the children after they had been using spans to measure. As they listened for the second time some of them were using their hands to help get an image of the actual measurements. There were some puzzled faces and exclamations because the numbers of hands and items mentioned in the rhyme did not match the children's expectations.

'Our door is more than four hands.'

'Was it the teacher's hands?'

The children just had to test this one out. They wanted time to find out the actual measurements of similar items using their own hands as measures. At report back it was decided that a giant's hand must have been used.

Seeing the need for standard measures

As the children reported back, they discovered that there were variations in their measurements. The children suggested comparing hand sizes. Hand prints were suggested, and it was soon discovered that there was a great variation in the span of the prints. Attempts at ordering by size resulted in a pictograph showing the most common size. The children suggested that these 'common' prints be used for all future measuring.

Estimating

Now a 'common' measure had been arrived at the children wanted to measure all over again. We suggested that they should estimate first and then measure.

Improvising

The children wanted to fix up this rhyme so that it worked for their 'hand' before going on to change it, using objects and people that they wanted to measure. As we were making the final drafts of the children's rhymes it was suggested by one child that we leave spaces for the numbers so that the book could be given to another class to estimate how many hands. The answers were given at the back in case the other children didn't want to measure.

Extending to area

Since we had the hand prints available we improvised on the rhyme, changing the second line to,

> 'Four hands is what I need to cover the book'

and so on. This led to informal measurement of area and the idea of covering a surface. Further improvisations were then made.

Other measuring rhymes

The following rhymes can all be used to introduce, consolidate or extend the children's measurement concepts. The rhyme 'Measuring' for instance can be used to introduce measuring with informal units or to introduce more complex measuring situations such as,

'How do you measure a house, an elephant or a mouse?'

A piece of string

The class is split into two groups. The first group says the rhyme; the second group clicks their fingers and says 'Yeah'.

Today when I was playing
I found a piece of string. Yeah
It was just the right length
To use to measure things. Yeah

First I measured Daddy
He was ten strings tall. Yeah
Mummy was eight and a bit strings,
Then I measured the wall. Yeah

Four, five I counted
As I measured the window sill. Yeah
I wish I could measure our baby
But he won't keep still. Yeah

I'd like to measure an elephant,
Or even a little mouse, Yeah
I wonder how many strings it would be
To the very top of our house? Yeah

We found that older children were very creative in response to the last line of this rhyme and made suggestions such as,

'Sit on the roof and throw a ball of string down while you hold the top, then cut it into little strings.'

'Measure how many bricks in one string, then count.'

'A house is about six of me so I lie down, we put bits of string beside me six times then count.'

How many, how far?

Six groups take it in turns to say a verse of the rhyme.

How many steps do I have to take
To get from here to the door?
Please count how many steps I take
And tell me how many more.

How many squares will I have to use
In a path from here to the door?
Please count how many squares I use
And tell me how many more.

How many cubes will I have to use
To fill this toothpaste box?
Please count how many cubes I use
And tell me how many more.

How many blocks will I have to use
For a tower as tall as my desk?
Please count how many blocks I use
And tell me how many more.

How many bricks will I need to use
To balance this toy car?
Please count how many bricks I use
And tell me how many more.

'How many, how far?' extends the measurement to informal exploration of distance, mass, area and volume and invites the children to estimate and then complete in informal measurement units.

Before reading this rhyme to the children we prepared resources to match the measuring in each verse so that work stations could be set up. The children explored the ideas introduced in the rhyme before making up similar verses of their own, collecting objects of their own choosing to explore.

What shape is water?

Water in bottles,
Water in pans,
Water in kettles,
Water in cans —
It's always the shape
Of whatever it's in,
Bucket or kettle,
Or bottle or tin.

We used this rhyme to introduce informal explorations of capacity. After reading the rhyme to the children, and discussing what they had already observed about water, we provided a variety of containers for them to explore. They observed what happened as they poured from one container to another and began to make predictions about where the water would go and how full it would be. We asked the older children to predict and mark on the containers what would happen to the water if they tipped the container on its side or upside down (with the lid on).

This informal exploration generated questions like

'Which container holds most/least?'

'How many of these to fill this one?'

Time

Dinner

Three children say the middle lines with all the class saying 'Yeah!'.

All	Dinner is often ready at quite the wrong time.	Yeah!
Child 1	When you're in the middle of a game,	Yeah!
Child 2	When you're watching a really good show on TV,	Yeah!
Child 3	When you've just come across something *fascinating* outside,	Yeah!
Child 1-3	You'll find that dinner will be ready in half a minute.	Yeah!
All	But when you're STARVING, and the whole house smells of roast chicken — *then* dinner won't be ready for another hour.	Oooh!

Activities

We shared this rhyme with the children and asked if dinner was usually ready at the wrong time in their house. We discussed the various times that children had their dinner and when they watched television. This provided a good opportunity to show some o'clocks and half pasts on the clocks.

Recording

The children wrote about what they did after school and at what time. We used these to generate discussions and comparisons, e.g.

> 'I have my dinner half an hour after Jane.'

and to show these on clock faces. Some children counted the minutes in fives as they moved the hands around the geared clocks.

Although we didn't improvise on the actual rhyme the children made up rhymes of their own which we all vocalised. One example was:

Child	**What do you do at 3 o'clock?**
Group 1	**3 o'clock, 3 o'clock, we run home from school.**
Child	**What do you do at 4 o'clock?**
Group 2	**4 o'clock, 4 o'clock, we watch the Ninja Turtles.**
Child	**What do you do at 5 o'clock?**
Group 3	**5 o'clock, 5 o'clock, we learn our words.**
Child	**What do you do at 6 o'clock?**
Group 1	**6 o'clock, 6 o'clock, we eat our tea.**
Child	**What do you do at 7 o'clock?**
Group 2	**7 o'clock, 7 o'clock, we have our shower.**
Child	**What do you do at 8 o'clock?**
Group 3	**8 o'clock, 8 o'clock, we're snoring well.**

Other time rhymes

Dandelion clocks

**Wh - wh - wh - wh!
One o'clock, two o'clock, three o'clock, four,
I found a fairy close to my door.**

**Wh - wh - wh - wh!
Five o'clock, six o'clock, seven o'clock, eight,
I blew and I blew and I found it was late.**

**Wh - wh - wh - wh!
I blew and I blew 'til I counted to ten,
And now I'm beginning all over again.**

Wh - wh - wh - wh!

After learning this rhyme, many of the children asked,

'Do dandelion clocks really work?'

We decided to go outside and find out. The children recorded the different times that each dandelion clock showed so that we could compare results back in the classroom.

Recording

Each group of children was asked to sort out their own collected information and to find their own way of showing the results of their research to the rest of the class. This resulted in a variety of recording methods including stories, lists and pictographs. Some children showed the time on digital or analogue clocks while others tried writing the times in words.

This generated a lot of interest in different ways of saying and recording times.

Bathtime

A group of children, with a girl and a boy to play the Mum and Dad.

Group   **When's breakfast ready Mum?**
Child 1   **At half past seven.**

Group   **When's lunch ready Mum?**
Child 1   **At half past twelve.**

Group   **When's dinner ready Mum?**
Child 1   **When you've had your bath.**

Group   **When's bathtime Mum?**
Child 2   **Now said Dad, and grabbed them from behind.**

An important part of learning to tell the time is sequencing events and beginning to relate times to special events in the day. We read 'Bathtime' to the children and introduced the term 'half past' inviting the children to suggest how half past might be shown on a clockface and why. The children wanted to add extra lines to this rhyme and make it include other events that were important to them. Some children found out the actual times of events for their own families and many improvisations of the rhyme were made and illustrated.

Hickory dickory dock

The children sing:

Hickory dickory dock
The mouse ran up the clock.
The clock struck one,
The mouse ran down,
Hickory dickory dock
Tick, tock, tick, tock.

The children suggested this rhyme because they knew other versions of it and thought we could improvise on it. Some examples included,

The clock struck eight
The mouse was late . . .

The clock struck six-thirty
The mouse was dirty . . .

The improvisations were illustrated with times and sequenced before being stapled into a class book.

Come on Dad

A group of children, with a boy and girl to play the Dad and Mum.

Group	**Come on Dad.**
Child 1	**Wait a minute.**
Group	**Come on Mum.**
Child 2	**Won't be long.**
Group	**Soon Dad.**
Child 1	**Wait a minute.**
Group	**Now Mum.**
Child 2	**Won't be long.**
Group	**How much longer must we wait?**

What does 'won't be long' really mean? Is a minute always a minute or sometimes more? We used this rhyme to explore the passage of time; what does a minute feel like, what can we do in a minute, and so on.

The ten o'clock scholar

A diller, a dollar, a ten o'clock scholar!
What makes you come so soon?
You used to come at ten o'clock,
But now you come at noon.

After sharing this rhyme with the children we asked them what they thought it meant. This required an explanation of the term 'noon' and generated discussion of other time words. Midnight was know to all, midday to some. A list of words including breakfast time, lunchtime, dinnertime, noon, mid-day, afternoon, morning, night time, evening, dusk, dawn, sundown, sun-up, bedtime, elevenses was generated. The children provided illustrations and clock times for many of these and arranged them on a time line.

Riddles

Riddles are a lot of fun and can just fill that odd moment or set the lesson off to a fun start. We've found that children learn from riddles too in their desire to 'get the joke' and to be able to remember and retell the riddles or to embellish and improvise on them.

A star

Higher than a house, higher than a tree.
Oh! whatever can that be?

Getting up

Question: **Why is getting up at three o'clock in the morning like a pig's tail?**
Answer: **It's twirly.**

Two's company

Question **If two's company and three's a crowd, what's four and five?**
Answer **Nine.**

10 cents

Teacher	If you had 10 cents, and you asked your dad for another 10 cents, how much would you have?
Boy	Er . . . 10 cents, sir.
Teacher	You don't know your arithmetic, boy!
Boy	You don't know my dad, sir.

Ten oranges

Teacher	If I had ten oranges in one hand, and seven in the other, what would I have?
Girl	Big hands, miss!

99 bonk

Question	What goes 99 bonk?
Answer	A centipede with a wooden leg.

Three eggs

Child 1	Did you hear the joke about the three eggs?
Child 2	No?
Child 1	Two bad.

Rabbits

Teacher	If I gave you three rabbits, then the next day I gave you five rabbits, how many would you have?
Girl	Nine, miss.
Teacher	Nine?
Girl	Yes, miss. I've got one already.

Time flies

Child 1 Why did the boy throw the clock out of the window?
Child 2 To see time fly.

Twelve months

Child 1 What happened to the man who stole a calendar?
Child 2 He got twelve months.

Yesterday's soup

A tramp went to a lady's house for something to eat.
'Would you mind eating yesterday's soup?' said the lady.
'No,' said the tramp.
'Good!' said the lady. 'In that case, come back tomorrow.'

Five eyes

In Hans' old mill his three black cats
Watch the bins for thieving rats.
Whisker and claw, they crouch in the night,
Their five eyes smouldering green and bright.
Squeaks from the flour sacks, squeaks from where
The cold wind stirs on the empty stair,
Squeaking and scampering, everywhere.
Then down they pounce, now in, now out,
At whisking tail, and sniffing snout,
While lean old Hans he snores away
Till peep of light at break of day.
Then up he climbs to his creaking mill,
Out come his cats all grey with meal —
Jekkel, and Jessup, and one-eyed Jill.

MIDDLE PRIMARY

7 YEARS AND UPWARDS

Counting

Mathematics is a cultural and a social process. Each society developed its own counting system to meet its own needs. Children understand their own number system, and maths in general, better when they begin to see that maths, like language, has developed and changed to meet the changing needs of society, and that it will continue to change in the future to meet the needs of a technological society. We used the rhymes and stories in this section to generate curiosity about numbers, place value and different bases.

Yan, tan, tether

Yan, tan, tether, mether, pimp.
Sether, hether, hother, dother, dick.
Yan dick, tan dick, tether dick, mether dick, bumfit.
Yan bumfit, tan bumfit, tether bumfit, mether bumfit, gigot.

Activities

We gave the children a well-spaced copy of this rhyme, without telling them that it was a counting rhyme, and asked them to read it and see if they could work out what it was. They came to the conclusion that it *was* a counting rhyme quickly, but couldn't immediately explain how it worked. We told the children that they could write or draw in the spaces between the lines if they thought it would help. This was the trigger the children needed. Some simply wrote the equivalent numbers underneath each word, e.g.

yan dick tan dick
11 12

and were still stuck. They were asked to explain why 'yan dick' meant eleven and so on. This helped them to see that the counting pattern wasn't quite that simple.

Some children wrote the number equivalent for each separate word, e.g.

yan dick yan bumfit
1 10 1 15

and so could see that 'yan dick' was 11 and 'yan bumfit' was 16.

We asked the children to continue the counting as far as they could. Some invented their own words for subsequent tens.

Sharing

At the report-back session we asked the children to explain how the counting system worked and how it was the same as/different from our system. We also asked them to comment on which was most useful and why.

Improvising

The children used objects and place-value charts to explore other ways of grouping and then made up their own counting system to match. They felt that the grouping in fives in the original rhyme gave it away a bit and so decided to use rhythm or arbitrary line length for their own rhymes. The rhymes were then used as puzzles, each group trying to work out how the number system worked.

Roman numerals

Some children were familiar with roman numerals and wanted to find out how they worked as a counting system. When the children were confidently reading and writing roman numerals we asked them why they thought they weren't widely used. We then set a few simple sums in roman numerals and the children were amazed that they were so difficult to use in that way.

Stories for counting

1. Counting sheep

This story involves the children in another problem of counting — ways of showing numbers using the fingers of their hands as the 'symbols'. As you read the story to the children, pause and allow time for them to try possible ways of showing numbers using the fingers of one hand.

Many years ago, on an island far away, the islanders all kept sheep. They had been doing very nicely thank you until one day a dragon came to live on the island. The dragon ate sheep. In fact, sheep were its favourite food. The islanders became very poor and tried hiding their sheep in caves and forests but the dragon always found them. The islanders wondered how he found them.

Now there is one thing you ought to know and that is that on this island when people meet on the street the polite greeting is not 'Hello, how are you?' but 'Hello how many sheep do you have?' The islanders loved to compare the size of their flocks.

One young islander who had spent a lot of time thinking about this problem suddenly realised how the dragon knew how to find the sheep. [Do you know?] He ran off to tell everyone.

'The dragon listens to you greet each other and then he follows the person with most sheep. You must stop telling each other how many sheep you have.'

Well, as you can imagine, the islanders did not like this at all. What could they do if they could not talk about the very

thing that interested them the most? The young islander had thought about that.

'None of you has more than five sheep left, so just use the fingers of one hand to show how many sheep you have.'

This worked well until after the lambing season when the islanders had more than five sheep each. They asked the young islander to help.

'You can still show how many sheep you have on the fingers of one hand,' he said, and explained what they must do.

'Use the little finger for one the second finger for two and so on.'

The villagers tried this and found that by using different combinations of fingers they could even show numbers as large as fifteen. The dragon was starving and could be found listening, listening but he never heard what he was listening for. As time went by they gave new numbers to the fingers of their right hands so that they could show even larger number. [Do you know what the largest number you can show is?]

Today the islanders are very rich and their flocks are so large that it takes the fingers of two hands to show how many sheep they have. And yes, the islanders still use their fingers for, althought the dragon has left the island, you never know, he might come back sometime.

Storytelling and participation

Telling this story, rather than reading it, allows for embellishments and a high level of involvement. Pausing before giving away how to count beyond nine does not get in the way of the story and really encourages the children to participate. The children can greet each other, showing a combination of fingers so their opposite numbers can try to work out how many sheep they have. No one seems to notice that number facts are being practised.

Investigating

The children were intrigued to find that by using just one hand they could make combinations to show all the numbers to thirty-one. One child announced that by using two hands they could show all the numbers to sixty-two. We asked the children if they agreed. Initial comments included,

> 'All the left-hand fingers could be tens so that's 81.'
>
> 'It must be 310 on the left then, 10, 20, 30, 40, 80 and 160.'
>
> 'You don't need ten or twenty you can do them on the right hand.'

We encouraged the children to use diagrams to help them work this out.

Recording

There were some differences in the numbering of the fingers of the left hand and these were discussed. It was decided by the children that each idea was valid so they drew diagrams of their right and left hands showing the value of each finger and posed problems to the rest of the class.

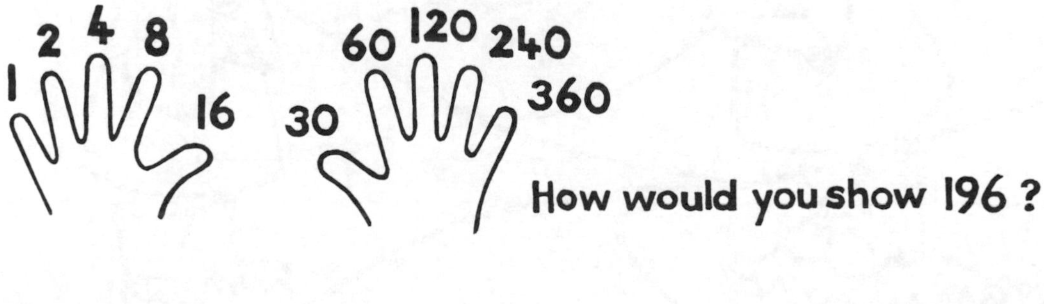

1 2 4 8 16

30 60 120 240 360

How would you show 196 ?

2. Inventing numbers (Fact)

Did you know that thousands of years ago, before numbers had been invented, people who wanted to keep a tally of their sheep or the number of days made notches in pieces of wood, or marks like lines in clay? Symbols were invented later to represent larger numbers. The Mayans, completely cut off from the rest of the world, developed a system where they could show any number they wanted using only three symbols. These were a dot, a line and an oval.

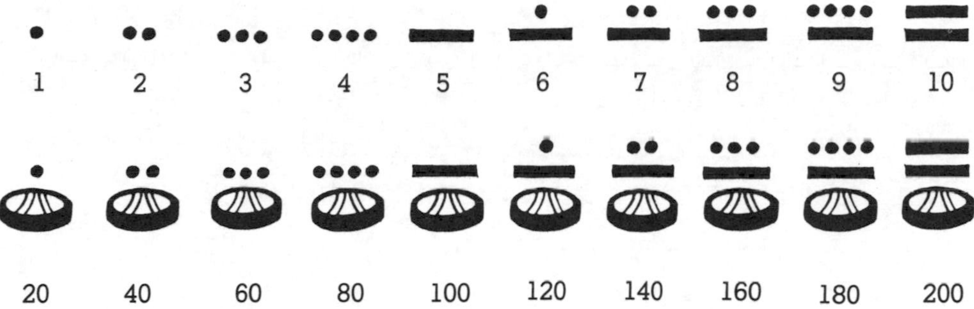

Investigating

We gave the children the example of the Mayan number system and asked them to find out how it worked. After a short time we asked the children what they thought the oval meant. Some thought that it meant something different every time. When we asked if they thought that it was good to have a symbol that could mean lots of things they said no, so we asked them what they thought the oval might be doing to each number if it was not adding 9, 19, 59 and so on.

The children soon worked out that the oval represented 'multiply by twenty' but wanted to know how to show 21 or 22. They investigated this themselves, some arriving at:

The children extended this idea by inventing their own rules for the oval. These were tried out on the rest of the class.

Research

Many children were so interested in numbers and counting systems that they wanted to research the history of numbers. A Chinese boy in the class came in with all the numbers 1–100 written in Chinese. We suggested that he record these in order onto a 100 grid which was then cut up into a puzzle. Several children who spoke a second language at home demonstrated and recorded how they count to 100 in their language.

Number facts

One old Oxford ox

The first group throws out a challenge that the second group replies to.

Group 1	Group 2
One old Oxford ox	**opening oysters.**
Two toads totally tired	**trying to trot to Tisbury.**
Three thick thumping tigers	**taking toast for tea.**
Four finicky fishermen	**fishing for finny fish.**
Five frippery Frenchmen	**foolishly fishing for frogs.**
Six sportsmen	**shooting snipe.**
Seven Severn salmon	**swallowing shrimps.**
Eight eminent Englishmen	**eagerly examining Europe.**
Nine nimble noblemen	**nibbling nectarines.**
Ten tinkering tinkers	**tinkering ten tin tinder-boxes.**
Eleven elephants	**elegantly equipped.**
Twelve typographical topographers	**typically translating types.**

Activities

Estimating and checking

We read this rhyme with the children and asked them to estimate how many creatures were in it. We then asked them what would be a good strategy for checking. Suggestions ranged from just counting on to spotting tens or known facts to using a calculator.

Reporting

At the report-back session the strategies were demonstrated, compared and discussed. The group who had spotted tens said they had found their answer more quickly than the people using the calculator.

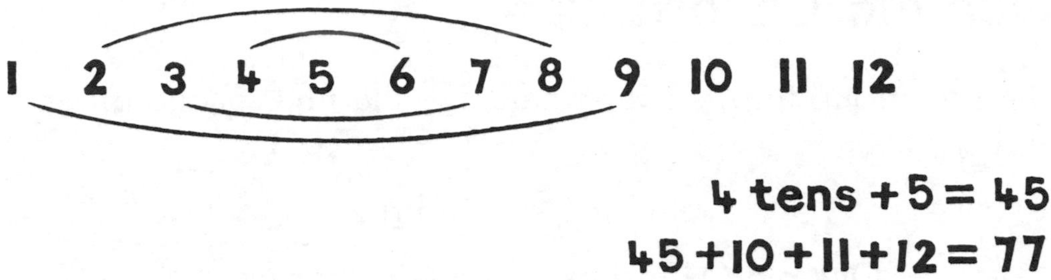

The calculator group said it could work out another similar sequence faster than the tens group. The numbers 3 to 15 were used this time and sure enough the calculator group was faster, but only because they had developed a strategy that combined grouping and the use of the calculator. When asked what they'd done they said,

> 'We just matched from top to bottom and found we'd got 6 lots of 18 and 9 left over so we did 6×18+9 on the calculator.'

3 4 5 6 7 8 9 10 11 12 13 14 15

$$6 \times 18 + 9 = 117$$

The other group had spotted the only 10 and then added on the remainder. The rest of the class now wanted to try these strategies and all recognised that grouping the numbers in the string was a good strategy. One group saw a relationship and arrived at a formula.

> 'You count how many numbers, then see what the top and bottom add to, then you halve the amount of numbers, multiply the answer and add on the left over.'

Investigating

The group explained and demonstrated this relationship to the rest of the class whose first comment was,

> 'Does it always work?'

The children were feeling quite confident now, so with the aid of their calculators they began to explore other combinations, and some with larger numbers too.

Improvising

The children decided to improvise on this rhyme to make it into a counting rhyme for younger children or a puzzle rhyme for older children. Some children used their tables to generate the numbers and were delighted to find that their formula still worked.

Other number fact rhymes

We used the rhyme 'One old Oxford ox' to introduce strategies for adding strings of consecutive numbers. Rhymes can provide opportunities to develop other strategies as well as offer an introduction or consolidation of a variety of number facts.

Bushrangers

All the class claps to the rhythm of the rhyme.

Fifty burly bushrangers
Went out to steal some gold,
But all the bush was wet with dew
And one caught a cold.
And one found a bulldog ant,
Creeping on his chest,
And one had a 'gammy' leg
And had to have a rest.
One thought he saw a snake,
Another had a pain,
The rest, they heard a gun go off,
And scampered home again.

We used this rhyme to consolidate the bridging of the tens, pausing at the end of each line for the children to predict how many would be left each time.

Improvising

The children improvised on this rhyme by varying the number of bushrangers to start with and/or the number that left each time.

> Sixty burly bushrangers
> Went out to steal some gold,
> But all the bush was wet with dew
> And two caught a cold.

When the children read their rhymes to the rest of the class, they too paused for the others to say how many were left. Some children realised that there were five occasions when bushrangers left, so combined 5×2 or 5×3, etc. with subtraction to predict how many would be left at the end.

Two in a boat

Two children sit on the floor, hands linked, and 'row' to the rhythm of the rhyme. After 'by and by' each child chooses a new partner before the next verse begins.

Two in a boat and the waves run high
Two in a boat and the waves run high
Two in a boat and the waves run high
Get me a partner by and by.

Four in a boat and the waves run high
Four in a boat and the waves run high
Four in a boat and the waves run high
Get me a partner by and by.

Eight in a boat and the waves run high
Eight in a boat and the waves run high
Eight in a boat and the waves run high
Get me a partner by and by.

Doubles

'Two in a boat' demonstrated doubling very well, and the children were then able to give explanations of the doubling process. We used this rhyme to extend the range of doubles that the children were confident with and to find out what they thought doubling was. We also found, like David Fielker in his research, that children are not sure what doubling is. When the lines were known to the children we used different starting numbers, 3, 5, 6 and so on.

The children noticed, as they dramatised the rhyme, that when numbers are doubled they grow very quickly and began to ask questions such as,

> 'If we start with 3/5/6/7 how many doubles before we run out of children?'

> 'If we double 3 ten times what might the answer be?'

The animals in the ark

The class sings this song together.

The animals went in one by one,
Hurrah, hurrah.
The animals went in one by one,
Hurrah, hurrah.
The animals went in one by one,
The elephant chewing a caraway bun,
And they all went into the ark
For to get out of the rain.

The animals went in two by two
The centipede and the kangaroo.

The animals went in three by three,
The pig, the rabbit, the chimpanzee.

The animals went in four by four,
The giant 'potamus got stuck in the door!

The animals went in five by five,
'Step along smartly! Look alive!

The animals went in six by six,
The monkeys were up to their usual tricks.

The animals went in seven by seven,
Said the ant to the antelope, 'Who are you shoving?'

The animals went in eight by eight,
The worm was early. The bird was late.

The animals went in nine by nine,
The giraffes stood straight and marched in line.

When the children were familiar with this song we paused at the end of each verse for them to predict how many went into the ark each time and then to use objects to show how the animals went in. The children had three different views of how the animals went in, so '3 by 3' was interpreted by some as one lot of three, by others as two lots of three and by still others as three lots of three. Reasons for these differences were discussed, then the children wanted to know how much difference the interpretations made to the answers.

From this the children developed an awareness of square numbers. They noticed that 1, 4, 9 . . . could be shown as squares and wanted to find out if any of the missing numbers could be shown as squares and also to explore what other higher numbers could be made. They also looked at the pattern in the square numbers and used this to predict others in the sequence.

Biscuits

Spoken by the class, with great emphasis on the rhythm.

Three little biscuits sitting on a plate,
Three for Molly, none for Kate.
None for Bill, none for his mate,
And none for the little boy who comes in late.
Yeah, yeah, yeah, yeah.

Six little biscuits sitting on a plate,
Three for Molly, three for Kate.
None for Bill, none for his mate,
And none for the little boy who comes in late.
Yeah, yeah, yeah, yeah.

Nine little biscuits sitting on a plate,
Three for Molly, three for Kate.
Three for Bill, but none for his mate,
And none for the little boy who comes in late.
Yeah, yeah, yeah, yeah.

Twelve little biscuits sitting on a plate,
Three for Molly, three for Kate.
Three for Bill, three for his mate,
But none for the little boy who comes in late.
Yeah, yeah, yeah, yeah.

Fifteen little biscuits sitting on a plate,
Three for Molly, three for Kate.
Three for Bill, three for his mate,
And three for the little boy who comes in late.
Yeah, yeah, yeah, yeah.

As we read this rhyme we paused for the children to predict who would get biscuits each time. After listening to the rhyme the children decided that it wasn't really fair that some people got three biscuits and some got none and that they should have been shared differently.

Modelling

We provided plasticine and Wainwright fraction kits, paper and scissors so that the children could investigate different methods of sharing. This led to some sophisticated exploration of fractions, and to factors as one group explained,

> 'We cut the biscuits into thirds, then the thirds into thirds but we never got a number that four could share. When we tried halves each time everyone got six pieces of one-eighth. We found that you have to cut into a number that four can share, like, 8, 12, 16, 20 . . . so 3s will never work, 5s might work though.'

Good morning girls and boys

One group plays the angry teacher, while the rest of the class chants their tables. While the group is performing, the class could repeat their lines *very* quietly, like a round.

Class **Once eight is eight.**
Two eights are sixteen.
Three eights are twenty-four
Four eights are thirty-two . . .

Group **Hands on heads and hands on hips,**
Now sit up in your seat,
Simon says to sit up straight,
Kindly move your feet.
Take your books out,
Rule a margin,
Keep it very neat,
Good morning girls and boys.

Stop that wriggling,
Stop that talking,
Get back on the track.
Stop that mumbling,
Please be quiet,
I'll give that boy a whack!

Don't do that please,
Stop that fiddling,
Hands behind your back,
Good morning girls and boys.

Class Five eights are forty
Six eights are forty-eight
Seven eights are fifty-six
Eight eights are sixty-four . . .

Group Walk in quietly,
Stand in twos,
This table's in a mess.
Don't do that please,
Put your pens down,
Don't say 'yeah', say 'yes'.
Stop that jiggling,
Put your hand down,
Wait until recess,
Good morning girls and boys.

Don't call out please,
Put your hand up,
Stay in after school.
Hands together, feet together
Please don't act the fool.
Wipe that mess up,
Fill this up please,
What's that little pool?
Good morning girls and boys.

Class Nine eights are seventy-two
Ten eights are eighty
Eleven eights are eighty-eight
Twelve eights are ninety-six . . .

Group **Clean your tables,**
See me later,
Bring a note from home.
Please don't use a calculator,
Stand here on your own.
Pull your socks up,
Take your things out,
Leave that girl alone,
Good morning girls and boys.

Pick that paper up please,
Now you know where paper goes.
Pick your bags up,
Pick that peel up,
Pick the best two rows.
Pick a captain, pick a team, but
Please don't pick your nose!
Good morning girls and boys.

We used this rhyme simply for fun and found that because the children enjoyed mimicking the grouchy teacher they wanted to learn it. It seems that learning the eight times table is not that difficult when it is a source of fun.

Time

Dinner

Three children say the middle lines with all the class saying 'Yeah!'.

All	**Dinner is often ready at quite the wrong time**	**Yeah!**
Child 1	**When you're in the middle of a game.**	**Yeah!**
Child 2	**When you're watching a really good show on TV.**	**Yeah!**
Child 3	**When you've just come across something *fascinating* outside,**	**Yeah!**
Child 1–3	**You'll find that dinner will be ready in half a minute.**	**Yeah!**
All	**But when you're STARVING, and the whole house smells of roast chicken — *then* dinner won't be ready for another hour.**	**Oooh!**

Note: We have repeated this rhyme from the previous section because, when we used it with older children, we got quite different reactions.

Activities

Exploring the passage of time

Children need many opportunities to explore the passage of time and use appropriate language if they are to develop an awareness and understanding of time. We used 'Dinner' to introduce time as an area for investigation. We shared the rhyme with the children and invited comments about it. The children associated very closely with the ideas presented and expressed a wonderment about the way time sometimes goes slowly and sometimes goes too quickly. We asked the children to think about why this might happen. Suggestions include,

> 'We could see what ten minutes feels like when you do different things like sitting still or reading a book or playing a game.'
>
> 'We should see if time goes slower when you are waiting to do something you do like or when it's something you do not like.'
>
> 'We could see what you can do in a given time.'

The children then set up their own experiments using timers and clocks. As they worked, new ideas and questions arose, for example the group investigating what ten minutes actually felt like realised that ten minutes was too long for sitting still so decided to begin by doing different activities and estimating how long they each took. To do this they recorded their starting times, did the activity, made their estimates and then invented their own methods, using the clock face, to work out the passage of time.

Recording

The children wrote reports about their findings, drawing clock faces to show starting and finishing times and explaining how to use a timer. The information was also used as the basis for problems that were made into a class book, for example,

'How many times do you think you can skip in two minutes?'

'How long do you think it takes to count to 100?'

Improvising

The children improvised on this rhyme to make it match their own experiences and added clockfaces to their own versions to show the appropriate times.

Other time rhymes

Seconds

Two groups of children recite this as question and answer.

Group 1 **How many seconds in a minute?**
Group 2 **Sixty, and no more, are in it.**
Group 1 **How many minutes in an hour?**
Group 2 **Sixty for both sun and shower.**
Group 1 **How many hours in a day?**
Group 2 **Twenty-four for work and play.**
Group 1 **How many days in a week?**
Group 2 **Seven, both to hear and speak.**

Time

Two groups of children recite this as question and answer.

Group 1 **Sixty seconds make a minute,**
Group 2 **Not a lot can be done in it!**
Group 1 **Sixty minutes make an hour,**
Group 2 **Time for sun and time for shower.**

Group 1	**Twenty-four hours in a day,**
Group 2	**Time for work and sleep and play**
Group 1	**Seven days make up a week,**
Group 2	**Time to listen, time to speak.**
Group 1	**Four weeks, and a month has gone**
Group 2	**Things go right and things go wrong!**
Group 1	**Twelve months pass and it's a year,**
Group 2	**Time for sadness, time for cheer.**

We used these rhymes to introduce the number of seconds in a minute and to consolidate minutes in an hour and hours in a day. The children had heard the term 'half a minute' and had used 30-second timers but had never put the two together. A discussion of time terms in general followed, along with questions that the children generated,

'Do you really count to 60 and it's 60 seconds?'

'How many seconds in 2, 5, 10, 15 minutes?'

'How many quarters of an hour in an hour/day?'

Thirty days has September

Thirty days has September,
April, June and November.
All the rest have thirty-one.
Not February — it's a different one.
It has twenty-eight. That's fine!
A leap year makes it twenty-nine.

We introduced this rhyme to help the children to remember how many days in each month and how many days in a year. The children used calendars to see what day their birthdays and other significant occasions were on and began to realise that four weeks from today wouldn't be the same date the next month.

The children began to use the rhyme to help them predict what the date would be one month after any given date,

> 'Monday, January the 7th, 31 days in January, so that's Monday plus 3 days. Thursday.'

December

Twelve months in the year I know,
This is the way the twelve months go:
January, February, March, April,
May, June, July, August and still
to come — September, October, November
and last of all the best month, December.

We used 'December' to help children remember the names and sequence of the months. The children all liked December but had other favourite months too. Discussion of birthdays and holidays and so on generated discussion of what each month is like in terms of weather and activities and so a picture of the seasons was drawn up.

Probability

I never win at parties

The teacher pauses at the end of each line while the class echoes in a whisper.

I never win at parties.
I never win at all.
Someone gets the prizes,
Someone wins the ball.

Someone gets the roses
Off the birthday cake.
I don't get the roses,
I get the stomach ache.

Someone pins the tail
On the donkey's seat.
When I pin the donkey,
It ends up on his feet.

Someone drops the clothespins
Right where they should go.
I can't hit the bottle,
Even bending low.

I do not know the reason,
Unless it's that I'm small,
Why I don't win at parties.
I just don't win at all.

Activities

Children often make statements such as,

> 'I always lose at . . . '
> 'You always win.'
> 'It's never fair.'
> 'It always rains at weekends.'

We used this rhyme to set the children talking about fair/unfair, likely/unlikely events. After hearing the rhyme the children began to talk about their own experiences, so we asked questions to focus on the probability aspect of events,

> 'Why do you think you always win/lose?'
>
> 'What would have to happen to make you win/lose?'
>
> 'If you played 20, even 100 games would you still always win/lose?'

When interest was high we asked the children to mark on a grid how they thought they would go on some readily available games.

Will you win at:	unlikely	likely	very likely	certain
noughts and crosses				
heads and tails				

The children then tested their predictions by playing each game once. As we talked about the results, the children began to raise points such as,

> 'If we played different people we might get different results. We need to try it on lots of people.'
>
> 'We might win once but lose next time. It might not always be the same.'

'In some sports it's the best of three, so we should play each game three times to get a fair result.'

The children set up their own experiments, recorded the data they collected and used this to report to the rest of the class. They also began to be less categorical in their predictions.

Recording

The children recorded their data on grids, as tallies and as fractions and wrote reports using pictographs or numbers to show the information collected.

Improvising

The children improvised on the rhyme to make it match their own interests or experiences.

Other probability rhymes

What are your chances?

Three groups read successive verses.

Group 1 **When you spin a spinner,**
Which number is best?
Will one come up
More than all the rest?

Group 2 **When you toss a coin,**
Which side do you call?
Is 'heads' or 'tails'
Most likely to fall?

Group 3 **Is it all just luck?**
Is it nothing but chance?
Or is there a way
To tell in advance?

Investigating

When the children had recited this rhyme aloud, we asked them to suggest answers to the questions asked. They wanted to explore these questions as well as some of their own. As they experimented they began to see that they needed to find ways of recording the results each time so that they could look for trends or patterns.

The children used their own questions to improvise on the rhyme. The issue of luck and chance was not resolved although statements of likelihood were being made, for example,

> 'If there are ten numbers on the spinner there's less chance of spinning a one than if there are six numbers on the spinner.'

Red sky at night

Red sky at night,
Shepherds' delight.
Red sky in the morning,
Shepherds' warning.

We asked the children if this rhyme was true or false and how they could find out. Checking that evening and the next day's morning against tomorrow's weather was suggested but, as with the spinners, some children thought that it needed to be tested more than once. A longer term observation was decided upon.

The north wind doth blow

The north wind doth blow,
And we shall have snow,
And what will the robin do then?
Poor thing!

He'll sit in the barn,
And keep himself warm,
And hid his head under his wing.
Poor thing!

This traditional rhyme is clearly targetted at an audience in the northern hemisphere, where snow is a common occurrence. We live in Queensland and so discussed with the children what changes we could make. It was agreed that the following variation would suit our location.

The north wind brings a gale,
Then we shall have hail,
And what will the magpie do then?
Poor thing.
He'll hide his head under his wing.
Poor thing.

Dipping rhyme

One child says the chant while the rest stand in a circle.

Eeny, meeny, miny, mo,
Catch a tiddler by his toe.
If he hollers let him go,
Eeny, meeny, miny, mo.
(If you don't like your choice, add:)
O—U—T spells 'out',
So out you must go!
(And if you are still not happy, add)
Because the King and Queen say so.

For generations dipping rhymes have been popular as a way of choosing. We asked the children if this particular rhyme was a fair way of choosing. They decided to investigate the effects of different starting positions and soon discovered that it was possible to know where to start the rhyme in order to eliminate someone or to make sure that the person you want is chosen. They could also explain how it happened and concluded that if you didn't know about this the dipping method was fair otherwise it was unfair. To reward their efforts, we read them the following puzzle as a challenge.

Catching the mice

'Play fair!' said the mice. 'You know the rules of the game.'

'Yes, I know the rules,' said the cat. 'I've got to go round and round the circle, in the direction you are looking, and eat every thirteenth mouse, but I must keep the white mouse for a titbit at the finish. Thirteen is an unlucky number, but I will do my best to oblige you.'

'Hurry up, then!' shouted the mice.

'Give a fellow time to think,' said the cat. 'I don't know which of you to start at. I must figure it out.'

While the cat was working out the puzzle he fell asleep and, the spell being thus broken, the mice returned home in safety.

At which mouse should the cat have started the count in order that the white mouse should be the last to be eaten?

Space and measurement

The farmer's field

A long time ago in a far off land there lived a poor farmer who had just one square field. Each side of his field was exactly one hundred paces long. One day a weak and hungry beggar knocked at his door. The farmer fed the beggar and gave him a place to sleep. In the morning the beggar went on his way. That evening a grand procession advanced to the farmer's door. The beggar was dressed like a king, indeed he was the king of the land. He had come to repay the kindness that the farmer had shown him.

'I want to reward you for being so kind,' said the King. 'You may now make your field twice as large but you must still keep it square.'

The farmer was very happy and quickly worked out how to extend his farm. 'This is easy!' he said to himself as he measured two hundred paces from the corner of his field. He turned right and continued measuring. Soon he had marked out a square field for which each side was two hundred paces long.

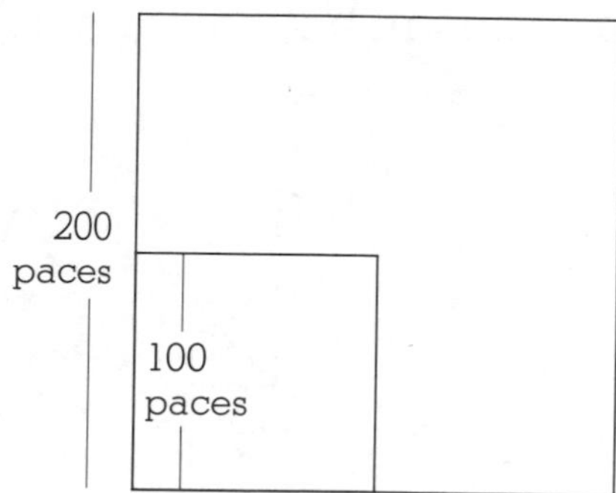

Pleased with his work, the farmer went to the king and explained how he had successfully marked out the

new field. The king drew the farmer's solution in the sand with a stick, and added more lines to the diagram.

'You greedy farmer!' said the king. 'You have made a field not twice as big but four times as big. If you cannot mark out a field that is only twice as big as the field you own, I will take back my reward, and I will confiscate your own field into the bargain.'

Activities

We read the children the story 'The farmer's field' and challenged them to write their own conclusion, encouraging them to continue the narrative in the style that had been started. Since the children invented many solutions we have listed a few possibilities below.

1. Mark in the diagonal of the square and make two more triangles the same. These pieces can be rearranged to make a new square twice the size of the original.

   + =

2. Mark in both diagonals, and fold out the four triangles to make a new square.

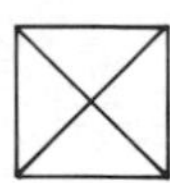

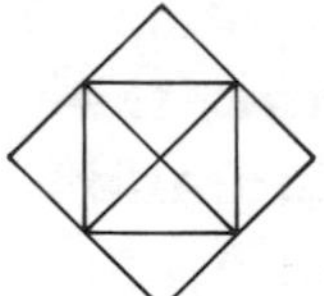

3. Measure the diagonal with a rope and use the rope as the length of the side of the new square.

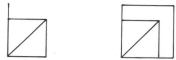

4. Measure 200 paces along the diagonal, and use this as the diagonal of the new square.

Most of the children used methods 1 and 2, but older children might well suggest methods 3 or 4. For them, a further challenge would be to show how to measure out a field 3, 4, 5, . . . times the size of the original.

Crossing the stream

During a country ramble, Mr and Mrs Softleigh found themselves in a pretty little dilemma. They had to cross a stream in a small boat which was capable of carrying only 80 kg [180 lb] weight. But Mr Softleigh and his wife each weighed 80 kg [180 lb], and each of their two sons weighed 40 kg [90 lb]. And then there was the dog, who could not be induced on any terms to swim. On the principle of 'ladies first' they sent Mrs Softleigh over; but this was a stupid oversight, because she had to come back with the boat again and nothing was gained by this operation.

How did they succeed in all getting across? You will find it much easier than the Softleigh family did, for their greatest enemy could not have called them a brilliant quartette and the dog was a perfect fool.

We found that the children really needed objects to help them solve this puzzle. They were surprised that so many (11) crossings were needed.

Riddles

Logic or reasoning tends to be neglected in the maths lesson. These riddles, as well as providing puzzles and fun, require different skills and creativity to solve. Use them as a 'warm up' at the beginning of a lesson, or as a 'wind down' at the end of a lesson. Either way, the children really want to get the joke or catch in the riddle, and the motivation and enjoyment that they generate can lead nicely into (or out of) the maths lesson. Occasionally, the joke or riddle may generate investigation and improvisations.

Legs

Two legs sat upon three legs
With one leg in his lap;
In comes four legs
And runs away with one leg;
Up jumps two legs,
Catches up three legs,
Throws it after four legs,
And makes him bring back one leg.

(A man sat on a three-legged stool with a leg of mutton on his lap. In comes a dog and runs away with the mutton. Up jumps the man, throws the stool at the dog, and makes him bring back the mutton.)

Fair shares

The fiddler and his wife
The piper and his mother,
Ate three half-cakes,
Three whole cakes,
And three-quarters of another.

The parson, his wife,
The clerk and his sister,
Went down to the meadow
All of a clister.
They found a bird's nest
With four eggs in it,
They took one each
And left one in it.

(If the fiddler's wife and the piper's mother are one and the same person the division of the cakes was not difficult. Likewise, if the clerk's sister has the happiness to be married to the parson there were only three bird's-nesters.)

Pairs or pears

Twelve pairs hanging high,
Twelve knights riding by,
Each knight took a pear,
And yet left a dozen there.

Jokes

Child	Did you hear about the well behaved little boy? Whenever he was good, his dad gave him 10 cents and a pat on the head. By the time he was sixteen, he had $786 and a flat head.
Child	Did you hear about the boy who was christened 6⅞? His dad picked the name out of a hat.
Teacher	If you had forty cents in one pocket and fifty-five cents in the other, what would you have?
Boy	Somebody else's trousers, sir.

Three tortoises

One day a big tortoise, a middle-sized tortoise and a little tortoise went into a cafe. They ordered three banana splits. While they were waiting, they noticed it had begun to rain.

'Look at that,' said the big tortoise. 'We should have brought our umbrella.'

'You're right,' said the middle tortoise. 'Let's send the little one back to get it.'

'I'll go,' said the little tortoise. 'Only you must promise not to eat my banana split while I'm away.'

Well, the big tortoise and the middle-sized tortoise did promise, and so the little tortoise set off. A few days later the big tortoise said to the middle tortoise, 'Come on, let's eat his banana split anyway.'

'All right then,' said the middle-sized tortoise.

At that moment, the little tortoise shouted from the back of the cafe, 'You do that, and I won't fetch the umbrella!'

Attitude towards maths

If we really want to know how confidently children are operating with mathematics, we need to encourage them to talk about what they like, dislike, can do, can't do, find easy, or find hard. The rhymes and riddles in this section can lead to just such informal discussions where the children feel able to express with honesty and in the security of a friendly group, exactly what their attitude to mathematics is.

Age

The children click (*) and clap (•) to the rhythm of the rhyme.

Mary's five and Bea is three
Bob is nine, that's three time Bea
In four more years I'll be eleven,
That's much better than being seven.
Grandad's sixty I was told,
How many years till I'm that old?
Strange how Mum's age never changes
She's been twenty-one for ages!

Activities

Word problems generate a sense of panic in some children so we used this 'Age' rhyme to see how the children would respond to it.

Calculating

We read the rhyme through once without pausing and then again with pauses for the children to actually work out how old the narrator is and how much older the grandad is. Since the children didn't seem to find this too difficult we asked them to comment on the rhyme and how they knew how to work out the ages.

Improvising

The children improvised on the rhyme to suit their own families or imaginary ones.

Comparing

Only then did we compare the rhyme with word number problems that are sometimes presented to them. This generated discussion of the sense of panic,

> 'What am I supposed to do?'
>
> 'Which bit am I supposed to use and which leave out?'

that word problems present. Because this was a rhyme the children had simply not associated it with word number problems. They were amused and thought there was a message there.

> 'Word problems may not be that bad after all.'

Other rhymes/riddles for attitude

Rather than outline activities for these rhyme and riddles we simply suggest that you read them with/to your students allowing time for discussion and free response.

Multiplication is vexation

Multiplication is vexation,
Division is as bad.
The rule of three doth puzzle me,
And practice drives me mad.

What do you get?

Teacher	If you add 387 and 769, then double it and divide by 5, what do you get?
Boy	The wrong answer, miss.

I think my teacher loves me

Boy	I think my teacher loves me.
Girl	How can you tell?
Boy	She keeps putting kisses by my sums.

Close your eyes

Close your eyes.
Think of a number.
Double it.
Add 12.
Double it again.
Take away 15.
Dark isn't it!

By inviting the children to improvise on the jokes, they can include items that they lack confidence in or that they want to highlight.

Integration

Soldiers

This story provides an opportunity for the children to use number, space and logic together to find out how the soldiers were arranged each night.

Long ago in Egypt the Pharaoh died. He was mummified, then set to rest in a sarcophagus that had nine rooms in three rows of three. The Pharaoh was placed in the centre room.

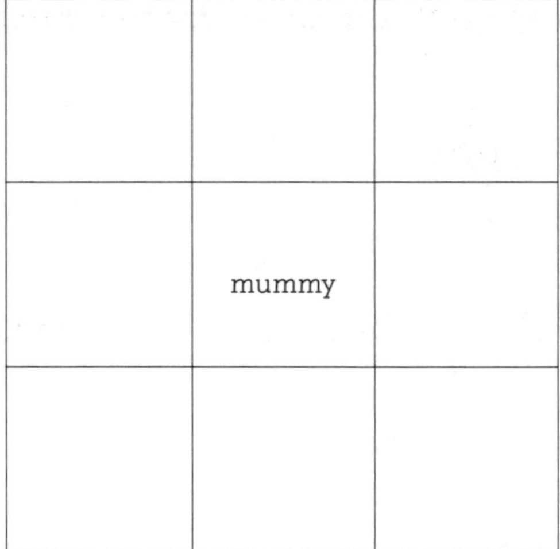

Sixteen soldiers were order to guard the tomb each night. At all times the soldiers were to make sure that six of them faced east, six west, six north and six south. And the first night this is exactly what they did.

Pause (How?)

The second night the soldiers invited two of their friends to join them, so eighteen soldiers guarded the Pharaoh. They were very careful though to make sure that six still faced east, six west, six north and six south.

Pause (How?)

The third night they invited yet two more guards to join them. So twenty soldiers guarded the Pharaoh. And of course for fear of angering the Pharaoh they had to find a way of arranging themselves so that six faced east, six west, six north and six south.

The fourth night the four visiting guards left and so did two of the soldiers. So only fourteen soldiers were left to guard the Pharaoh. They puzzled and puzzled and finally found a way of arranging themselves so that six faced east, six west, six north and six south.

Pause (How?)

We find it best if you read the story once without pausing and encourage the children to respond freely to it so that they can ask questions or make comments of interest to them. Then re-read the story, pausing as indicated, so that the children can use counters or diagrams to work out how the soldiers were arranged each time.

Note: You may prefer to memorise and embellish the story rather than read it to the children.

The children may want to find out if any other number of soldiers could have guarded the tomb. Others may want to try different sized sarcophaguses.

More ideas to inspire your teaching of mathematics.

Mathematics in Process
Ann and Johnny Baker

Mathematics in Process extends the confidence that teachers now feel in the language arts to the mathematics lesson. The purposes and conditions of natural learning, now common in the language classroom, are applied to learning and doing mathematics.

This very comprehensive book is divided into three parts.

Part One looks at the child's experience and has sections setting out how children get involved, how young mathematicians work, how children communicate and learn from reflection.

Part Two sets out classroom approaches: identifying purposes for using mathematics, the conditions for learning mathematics, shared experiences for learning mathematics and assessment.

Part Three relates ideas on devising a curriculum, how to set up the classroom and features a complete section on activities to try with your class.

Illustrated 176 pp

Maths in the Mind
Ann and Johnny Baker

Maths in the Mind focuses on the development of mental skills and strategies within the context of broader activities and emphasises the processes involved in mathematical thinking. Children are encouraged to explore, to delve into what they do know rather than feel helpless because of what they don't know. Twenty fully developed activities are set out in the book.

Maths in the Mind complements *Mathematics in Process*, also written by Ann and Johnny Baker.

Illustrated 120 pp

Maths in Context
Deidre Edwards

Deidre Edwards, an experienced teacher of primary school mathematics, shows how to integrate mathematics with the wider curriculum areas by using a central theme. As a result of this approach, she demonstrates that there is an increase in children's motivation, individual differences are catered for, children's confidence in their mathematical ability grows and mathematics is seen as part of 'real life'.

Maths in Context provides guidelines on how to handle group work, on classroom organisation and on planning and implementing assessment strategies.

A large section of the book presents ideas for activities based on the following themes: Dragons, Our Environment, The Zoo, Party Time, Traffic, Christmas, Show and Tell, The Faraway Tree.

Illustrated 152 pp

For further information about these and other exciting books for teachers contact:

Eleanor Curtain Publishing
2 Hazeldon Place
South Yarra 3141
Australia
(03) 826 8151

Heinemann Educational Books, Inc.
361 Hanover Street
Portland NH 03801 USA
(603) 431 7894

Ashton Scholastic
165 Marua Road
Auckland 6 New Zealand
(09) 696 089